AETHER-LIGHT

"THE FACT OF EVERYTHING"

Only one man knows in
the entire world.
1. Where did we come
from?
2. What is gravity?
3. What is matter?
4. What is dark matter?
5. What is dark energy?
6. What is aether?
7. What is light?

RANDY HOLMES

ISBN: Softcover   978-1-5245-7048-4
        Hardcover   978-1-5245-7047-7
        EBook   978-1-5245-7049-1

Print information available on the last page

Rev. date: 01/09/2017

To order additional copies of this book, contact:
Xlibris
1-888-795-4274
www.Xlibris.com
Orders@Xlibris.com

"Introduction"

"Aether Electromagnetic System" (AEMS) by: Randy Lee Holmes.

This book is about "Aether"; in physics proposed the existence of a "Medium" or a space-filling substance or field, thought to be necessary as a transmission medium for the propagation of electromagnetic or gravitational forces.

Since the development of special relativity by Einstein in 1905, theories using a substantial "Aether" fell out of use in modern physics.

Isaac Newton, James Clerk Maxwell, and Nikola Tesla, all believed electromagnetic waves of "Light" needed an "Aether" or "Medium" to propagate through empty space.

Since Einstein's special relativity in 1905, the scientific community recognizes the empty "Vacuum" for propagation of "Light" through empty space, and not "Aether".

Randy Lee Holmes has discovered visual proof and illustrations by more than 200 photos and 46 videos of "Electromagnetic Radiation" of the "Aether" that serves as a "Medium" for all "Light" and all "Electromagnetic Waves" of the universes.

My name is Randy Lee Holmes, and I have a BS degree in Physical Education, and I am a retired high school teacher of Cumberland County Schools in the state of North Carolina in 2016.

I have researched "Light" and "Anti-gravity" phenomenon for more than thirteen years (13) years, and now wish to publish this book on my discovery of "Aether".

I am proud of myself and thankful to God for this historic and great discovery that serves as my contribution to mankind.

This discovery of "Aether" will be recognized as man's greatest discovery in physics, the world, and the universe.

These are some of the categories where you may find my book.

Science, Physics, Light, Nature, Nonfiction, Futuristic, Photons, Gravity, Dark Matter, Dark Energy, Matter, Particle Physics, Theoretical Physics, Historical Discoveries in Physics and Science, and Nature.

As far back as 1678 scientist like Isaac Newton, James Clerk Maxwell, and Nikola Tesla, and the scientific community, all believed that "Light" needed a "Medium" to propagate through empty space.

It is a fact that sound waves (mechanical waves) require a material medium for its propagation such as gases, solids, and liquids.

The theory is that all waves need a "Medium" to propagate, as does electromagnetic waves must also need a "Medium" to propagate through empty space.

Since Einstein's special relativity in 1905, the scientific community recognizes the "Vacuum" for propagation of "Light", and not "Aether".

The modern theory is that "Aether" is not needed in order for "Light" to propagate through empty space, and only the "Vacuum" of space is needed for the propagation of "Light" and "Electromagnetic Waves" of the "Electromagnetic Spectrum".

The Electromagnetic energy that is attached to each individual photon is the "Aether" of the universe, and beyond.

"Light" can act as a particle because of the photon, and "Light" can act as a wave because of the Electromagnetic energy that is created and connected to the creation of each individual photon.

The materialization of Electromagnetic energy comes from the "Aether" of the universe, and beyond.

What many fail to understand is that the Electromagnetic Wave or "Aether Electromagnetic System" is the "Wheel-work" of energy in all the universe, and beyond.

The color bands of the triangle on your left are on the inside but the color bands of triangle on your right are on the outside.

This "Aether" photo was taken with a DC flashlight.

The "Aether Electromagnetic System" (AEMS) by Randy Lee Holmes 2016.

Notice that all stars in the Nebula are born with the "Aether Electromagnetic Radiation System Blueprint".

The "Aether Electromagnetic System" (AEMS) or "Electromagnetic" part of a photon serves as "Medium" through which the "Electromagnetic Wave" propagates.

"Light" provides its own "Medium" within the Electromagnetic Spectrum of Radiation; which is called the "Aether Electromagnetic System" (AEMS).

The "Aether Electromagnetic System" (AEMS) is the "Medium" of the "Electromagnetic Wave" that is woven into time and space.

The "Aether" of the Electromagnetic Spectrum consist of all Radiation in the entire universe, and beyond.

"Aether" is "Magnetism", which is always accompanied by "Electricity" within the Electromagnetic Spectrum of Radiation.

"Aether" of "Magnetism" is also responsible for all gravity of matter in the universe.

The Electromagnetic Spectrum of Radiation uses "Light" as nature's way of transferring energy through space.

Once man fully understands the "Aether Electromagnetic System" (AEMS) he will be able to travel through space faster than the speed of "Light".

The "Aether Electromagnetic System" (AEMS) embodies electrons, protons, neutrons, anti-matter, electromagnetic radiation, invisible light, visible light, transverse waves, radio waves, ultraviolet waves, infrared radiation, x-rays, gamma rays, matter, dark matter, dark energy, gravity, and all in the entire universe and more; as in other dimensions.

The "Aether Electromagnetic System" (AEMS) is the key to discovery of new worlds and other dimensions in our present and near future.

"Light" waves or photons are the "Electrical" component of the Electromagnetic Spectrum and "Magnetism" combines to form the "Aether" of the universe and transverse ways.

"Light" or photons are the energy that the four states of matter or all things are made of.

Every photon has the "Electromagnetic System Radiation Blueprint" of energy attached to it at the time of its creation.

Chemicals that activate into a toy glowing "Light" ball has the "Aether Electromagnetic System Radiation Blueprint " like all other "Light".

When a photon is confronted with the situation of whether or not to act as a particle or wave, it will use the "Electrical" part or photon part of itself to act like a particle, and "Magnetism" part of itself to act like a wave.

Some would explain it as the photon acting as a particle, and the Electromagnetic part acting as a wave; given the situation.

Maybe, the photon starts as a particle and builds itself into a wave.

The "Magnetism" is what holds the "Electrical Light" or photons, or "Matter" together to form the four states of matter.

The "Aether Electromagnetic System" (AEMS) is the "Medium" of the entire Electromagnetic Radiation Spectrum.

The "Aether Electromagnetic System" (AEMS) is the "Medium" for the propagation of "Light" and transverse waves.

Without the "Aether Electromagnetic System" (AEMS) there is no "Light" or "Electricity"; without "Electricity" there is no universe.

"Aether", "Matter", "Light", "Electricity", "Gravity", and "Space-Time", are all woven into the "Aether Electromagnetic System" (AEMS).

Space-Time and everything in the universe evolved from the "Aether"

and depend upon it for existence.

"Aether" started the creation of the universe; as it all started with "Light".

The photos in this book will verify these facts and the experiments done on the "Aether Electromagnetic System" (AEMS).

These photos shall prove that all "Light" of the Electromagnetic Spectrum need the "Aether Electromagnetic System" (AEMS) to propagate through space; whether it be AC, DC, flame "Light", or Star "Light" in the cosmos.

When you look at the photos in this book, you are looking at the "Electromagnetic Radiation Blueprint" of an actual star in the cosmos.

The "Aether Electromagnetic System" (AEMS) is the "Electromagnetic Radiation Blueprint" of all "Light" in the universe, and beyond.

"Photo Description and Proof of the (AEMS)".

Each beam of "Light" has its own "Turnstile" or "Crossed-Dipole Antenna" that serves as "Medium" of all Electromagnetic waves in the universe; and beyond.

These antennas are portals to other universes and other dimensions.

Notice the middle vertical line illustrates current and the horizontal Electromagnetic color band lines illustrates voltage.

The middle center of the vertical white "Light" line represents the source of "Light" being propagated and permeated through space.

Also, notice the horizontal electromagnetic Spectrum color bands are brighter on both sides of the vertical current white "Light" line in the center of the (AEMS).

It can be seen that the voltage is at a maximum at the center with presence of any current or source of "Light".

The different Electromagnetic Spectrum color bands on both horizontal lines of the "Aether electromagnetic System" (AEMS) indicates that the "Aether Electromagnetic System" (AEMS) does operate on any and all frequency.

The "Aether Electromagnetic System" (AEMS) serves as a "Transmission Medium" for all waves of the "Electromagnetic Spectrum".

"Aether Electromagnetic System" (AEMS) and more descriptions will follow.

When one looks at the "Aether Electromagnetic System" (AEMS) photos you will observe a white "Light" line of "Light" energy traveling down the middle of the "Aether Electromagnetic System" (AEMS) in the form of a white "Light" and vertical line from North to South.

This North and South vertical line of white "Light" is the "Light" or the "Electrical" component of the source of "Light" that is being propagated.

A second look at the "Aether Electromagnetic System" (AEMS) will have an x, +, or shape of the DNA chromosome of a human being; with all three shapes forming color bands of the Electromagnetic Spectrum on one or two crossing lines.

You can find these "Electromagnetic Spectrum" color band lines crossing to form an x, +, or shape of the DNA chromosome of a human being in all stars in the cosmos, and in all photons of "Light"; whether it be stars, filament, or candle "Light" flames.

According to James Hill and Barry Cox, when using Einstein's extended math equation it appears very clear that these x, +, and human DNA chromosome shaped lines are indications that these color band lines of "Light" are capable of traveling faster than the speed of "Light" (186,282 miles per seconds).

Einstein said nothing can travel faster than the speed of light of 186,282 miles per second or 299,792,458 meters per second.

These two East to West crossing lines of color bands will form a "Turnstile Antenna", two equilateral or reference triangles (x, +, or the shape of a human DNA chromosome) meeting at the center of the middle and vertical North to South white "Light" line.

A third look at the "Aether Electromagnetic System" (AEMS) and you will notice that the triangle on the left side of the "Aether Electromagnetic System" the color bands are on the inside of the triangle, but on the opposite and right side of the "Aether Electromagnetic System" (AEMS) the color bands are on the outside of the triangle.

The color band lines and triangles are the "Magnetism" component of the "Aether Electromagnetic System"(AEMS).

A fourth look at the "Aether Electromagnetic System" (AEMS) will show a perpendicular white "Light" line on most photos in this book.

The "Aether Electromagnetic System" (AEMS) incorporates the transverse wave, or the "Turnstile Radiation Antenna" to propagate "Light" throughout the universe; and beyond.

Before "The Big Bang" there was the "Aether Electromagnetic System" (AEMS).

God said "Let There Be Light".

This photo and all photos in this book are actual photos of the "Aether" and its electromagnetic radiation signature or fingerprint.

In simple terms, you are looking at the "Aether Electromagnetic System" in the form of radiation.

Notice that all photos of the "Aether Electromagnetic System" are exactly the same, and it does not matter what type of light you photograph.

Electromagnetic transverse waves travel 186,282 miles per second; it is what we are taught in science and physics class.

The "Aether Electromagnetic System" allows "Light" to travel faster than the speed of "Light"; as it serves to unify all nature in the universe; and beyond.

The "Aether Electromagnetic System" (AEMS) is an invisible and infinite material with no interaction with physical objects; for it is what all things are made of.

The "Aether Electromagnetic System" (AEMS) is massless mass without viscosity; much like the photon.

The "Aether Electromagnetic System" (AEMS) cannot affect the orbits of planets because it is the very wheel-work that makes the universe exist and function.

Static electricity "Light" or fluorescent bulb "Light" also have the "Aether Electromagnetic System" (AEMS) "Radiation Blueprint" when I am conducting my experiments on "Light"; and any type of "Light" has it.

The "Aether Electromagnetic System" (AEMS) is the "Medium" for propagation of "Light" in the form of Transverse Electromagnetic Waves; which can travel through dimensions.

Electrostatic "Light" does not travel by "Aether" at the present time, but it is my primary reason for living to find out how to perform such a task; because I know it to be possible to do using the "Aether Electromagnetic System" (AEMS).

Electrostatic charge is separation of charge or ions that I have not yet been able to combine with the photon.

My next discovery will consist of electrons accompanied by photons, and the discovery of transferring electrostatic charge by "Light" will bring about historic applications in anti-gravity, force fields, and limitless applications in anti-gravity technology.

All of this will be done with a beam of "Electrostatic Light".

The brightest light is from the flashbulb of the camera. The top light is from flame of torch.

All "light" carries the "Aether" and its radiation signature and fingerprint.

"Electromagnetic Waves"

The Electromagnetic Waves that compose Electromagnetic Radiation can be imagined as an "Aether" "Medium" propagating transverse oscillating waves of electric and magnetic fields of "Light" anywhere and everywhere in the universe, and beyond.

The "Aether Electromagnetic System" (AEMS) is a constantly moving of bodies that fuels oscillation of "Light", acting as the infinite wheel-work of electrical and magnetic energy for the entire universe, and beyond.

The "Aether Electromagnetic System" (AEMS) is the unlimited electrical and magnetic energy workings of the universe and space and time.

This "Aether Electromagnetic System" (AEMS) is always the same regardless of frequency or intensity of "Light".

The "Aether Electromagnetic System" (AEMS) is a "Devine Creation System" that was created by God when He said "Let there be Light".

Notice the middle and "Light" line is pointing to the right of the screen because I am standing closer to the right of the screen.

The "Light" beam will move from left to right, but the "Aether" signature or fingerprint will never move when up close.

If you back up ten or fifteen feet and move your head, your eyes will see the

"Aether Electromagnetic System" move slightly like an octopus with tentacles or waves with somewhat of snaking motion from side to side.

"Aether Electromagnetic System" (AEMS) Photos".

The "Aether Electromagnetic System" (AEMS) is an "Electromagnetic Radiation System Blueprint" that is visible to see with the human eye in the form of a photo or video; thanks to this book.

The "Aether Electromagnetic System" (AEMS) photos incorporates a transverse wave, or "Turnstile Radiation Antenna" propagating and oscillating photons of "Light" through space and time, and beyond.

A transverse wave is a moving wave that consist of oscillations occurring perpendicular to the direction of energy transfer.

If a transverse wave is moving in the positive x-direction, its oscillations are in up and down directions that lie in y-z plane.

The "Aether Electromagnetic System" (AEMS) incorporates the transverses wave in its "Aether" system.

All "Matter" is made of electrical massless "Light" (EMR), or photons.

The "Aether Electromagnetic System" (AEMS) photos in this book are evidence that the Electromagnetic Radiation of the universe is also the "Aether" of the universe.

The "Aether Electromagnetic System" (AEMS) is the "Medium" that allows Electromagnetic waves to oscillate and propagate.

The photos clearly shows "Light" as the vertical "Electrical" part of the system and the horizontal color bands and lines of Radiation as the "Magnetism" part of the "Aether Electromagnetic System" (AEMS).

Note that this "Light" line is pointing to the left, because I am standing to the close left of the "Light"

"Photons are Time and Space or Space-Time".

Every source of "Light" emits large numbers of tiny particles known as photons in a "Medium" surrounding the source.

Photons are massless "Light" of energy that all "Matter" is made of.

"Matter" is electrical energy according to $E=Mc2$; and "Matter" and energy are one in the same.

"Matter" and photons are also time and space in the quantum.

The "Aether Electromagnetic System" (AEMS) is a Radiation driven system.

Electromagnetic Radiation (EMR) is the "Light" that makes up the four states of "Matter" of solid, liquid, gas, and plasma.

Photons are massless energy of "Electromagnetic Radiation (EMR); and therefore, they are Electromagnetic Radiation of "Light".

Photons can perform as waves and massless photons; but photons can also appear to have mass and perform as particles; thus the "Photoelectric Effect".

Once man discovers how and why the Electromagnetic transverse wave is made to oscillate and propagate by the "Aether Electromagnetic System" (AEMS), he will discover the wheel-workings of the entire universe.

"Oscillation"

Massless "Light" cause electrons in a TV to oscillate and this oscillation generates an oscillating electric current.

When an oscillating voltage is applied to the TV transmitter, the electrons oscillates in response and the oscillation generates an Electromagnetic Wave.

"Light" is massless, but it is also an electric photon of energy in nature.

The "Photovoltaic" or "Photoelectric Effect" is the creation of voltage or electric current in a material upon exposure to "Light".

Top light is DC flashbulb from camera. Bottom photo is from candle flame.

"Light" or "Photon"

"Light" can act as a particle or a wave and it travels 186,282 miles per second is what we are taught in science and physics class.

One can only teach what one knows; or what he thinks he knows.

Is this scenario a particle or a wave, or both?

"Light" makes up the four states of "Matter"; and yet, "Light" is not "Matter"; or is it?

Yes, photons or electrical "Light" is what all "Matter" is made of.

In the beginning God said, "Let There Be Light".

"Light" can be in the form of "Matter" (Chair), or "Light" can be in the form of a massless photon of electrical energy.

Like "Light" is a massless electrical energy, then also is the "Aether Electromagnetic System" (AEMS)" a massless "Electrical" and "Magnetism" of energy, and more as a system.

"Light" or each individual photon has an "Electromagnetic Radiation Blueprint" of "Aether" that is shown in the form of "Radiation Blueprints" in all photos throughout this book.

Where there is mass there is gravity, and where there is "Light" there is "Aether".

There is no mass without gravity and there is no "Light" without "Aether" (By Randy Lee Holmes 2003).

The photos in this book are camera taken photos of "Light" and the "Electromagnetic Radiation Blueprint" of the "Aether Electromagnetic System" (AEMS).

Some photos were taken of AC "Light" (Lightbulbs in living room and kitchen). Some photos were taken of DC "Light" (Photos taken of "Light" with a flashlight).

Some photos were taken of flames of fire with a candle, match flame, cigarette lighter or torch.

You can find this same "Electromagnetic Radiation Blueprint" of "Aether" in all photos taken of any kind of Electromagnetic "Light" including the stars of the cosmos; and in any other form of "Light" of the "Electromagnetic Spectrum".

All photons or forms of "Light" have the exact same "Aether Electromagnetic Radiation Blueprint" that is in each photo in this book.

These photos in this book of the "Aether Electromagnetic System" (AEMS) has all visible "Light" colors of the "Electromagnetic Spectrum, and some white "Light", as well.

Notice the perpendicular line in this photo of the "Aether Electromagnetic System" (AEMS).

The "Aether Electromagnetic System" is Gods most beautiful creation in the world and the universe!!!

The Nebula is the most beautiful in the universe.

Notice that all stars in the Nebula are born with the "Aether Electromagnetic Radiation System Blueprint".

God created this "Aether System" when He said "Let there be Light" in the beginning of creation of all things and the universe.

"Description of Photo of Light".

1. All photos have a vertical straight white "Light" line traveling down the center of the "Aether Electromagnetic System" (AEMS) from North to South as a vertical line of white "Light".

2. All photos have two equilateral or reference triangles forming an x, +, or the shape of the DNA chromosome of a human being, and where the x or + comes together is the center where you find the straight white line down the middle of the "Aether Electromagnetic System" (AEMS).

The x, +, and shape of the human DNA chromosome lines are both lined with color bands of all the visible colors of the "Electromagnetic Spectrum".

These two lines that cross to make the x, +, or the human chromosome shape lines have all the different colors of the "Electromagnetic Spectrum" incorporated in both lines as far as the eye can see; they expand throughout the universe, and beyond.

These color bands are brighter towards the center white "Light" line traveling down the center of the "Aether Electromagnetic System" (AEMS).

3. The point or front point of both triangles meet at the center of the vertical white "Light" line of the "Aether Electromagnetic System" (AEMS).

The color bands are inside of the triangle on the left, but the color bands are outside of the triangle on the right; this means that one side is positive and the other side is negative, or one side is East and one side is West.

4. I feel that the center white "Light" line represents the "Electrical Light" part of the "Aether Electromagnetic System" (AEMS).

And both color band lines and both triangles represents the "Magnetism" part of the "Aether Electromagnetic System" (AEMS).

The "Magnetism" is an "Electrical Magnetism" that man is maybe unfamiliar with here on earth.

This "Electrical Magnetism" is the key to operations of the entire "Aether Electromagnetic System" (AEMS).

5. All photos show a spot of white "Light" at the top and at the bottom of the vertical white "Light" line traveling down the middle or center of the "Aether Electromagnetic System" (AEMS).

6. Most photos show a visible perpendicular line of white "Light" traveling from the top of the triangle on the left side through center white "Light" line to extend under the triangle on the right side.

7. All stars in the universe have this same "Aether Electromagnetic Radiation Blueprint" found in this book, and in these photos.

8. The two Electromagnetic color band lines shaped like an x, +, or human DNA chromosome shaped color band lines extend to the ends of the universe, and to the end of time and space.

These two Electromagnetic color band lines meet at the center of a star in the cosmos, and form an x, +, or human DNA chromosome shape as they extend through time and space as the "Aether" of the universe.

These two lines or "Turnstile Antenna " serve as "Medium" for transverse waves of the "Electromagnetic Spectrum" (By Randy Lee Holmes 2003).

9. "Light" is an Electromagnetic transverse wave that permeates and propagates in the Electromagnetic Radiation (EMR) of "Aether" of the universe, and of time and space.

10. Here is another theory on why all photos of "Light" have these lines, according to James Hill and Barry Cox of the University of Adelaide, Australia.

They have extended Einstein's equations to show what would happen if faster-than-light travel were possible.

The Electromagnetic color band lines that cross each other in photos of stars and photos of "Light" in this book could very well be an illustration of "Electromagnetic Radiation (EMR) undergoing the transition of space-time and traveling faster than the speed of light.

Flame of torch on top and flashlight on the bottom.

"Applications"

Once man can understand and control the "Aether Electromagnetic System"(AEMS) applications are limitless.

1. "Matter" can be transported from one point to another using "Light" energy without traversing the physical space between them (Teleportation or Tele-transportation).

2. Anti-gravity transportation vehicles.

3. Anti-gravity buildings in the sky.

4. Electrical fields of anti-gravity or repelling matter force fields to protect humans or property.

5. Plasma or lightning weapons.

6. Travel to other dimensions.

7. Travel to other planets in seconds.

8. Anything the human mind can conceive can be made reality using the "Aether Electromagnetic System" (AEMS) technology.

The top is a torch and the bottom is a flashlight.

"All Human Beings Made Of Light"

Matter and energy are both the same according to E=Mc2.

The Higgs boson (God Particle) shows the "Aether Electromagnetic System" reference triangles in all of its photographs.

This is more proof of both the "Aether Electromagnetic System" and "Aether" existence.

People are made of "Matter" or "Energy" or "Photons" or "Light".

When smashing together two protons in a collision you get "Light".

There should never be a debate about everything being made of light, but, knowledge is like life; some have and some have not.

Photons are pulses of "Light that all living and nonliving things are made of.

Everything is made of "Light Energy" or photons on earth and throughout the universe from the quantum to the cosmos.

The human body is a walking particle of Electromagnetic Radiation or Electrical Energy.

When man finally evolve, he will manipulate physical objects simply by his thoughts of electrical energies.

Just like today man can transport sound electrically all over the world, he shall be able to transport himself simply by using his mental energies, and a device.

Also, just like today man can send his physical body to another planet using the transverse electromagnetic wave; but he does not know this.

Man shall one day be one with himself and the energies of the entire universe.

The discovery of the "Aether Electromagnetic System" will open our eyes to technology that the human mind dare to conceived before now.

The four states of "Matter" are made up of "Light" and "Aether" or "Electrical Light".

We are "Matter" and we are "Light" (We do Matter, and we are the "Light").

The bottom "Aether" signature is from the candle light and the top is from a flashbulb.

"Is "Matter" and "light" both the same?

"Matter" and Energy are one in the same if you know about E=Mc2.

The "Aether Electromagnetic System" describes the particles and forces that account for all physical phenomena, and beyond.

The "Aether Electromagnetic System" includes the four fundamental forces governing interactions between particles: 1. "Gravity", 2. "Electromagnetism" (which is responsible for "Light", "Magnetism, and "Electricity), and the 3.

"Strong" and 4. "Weak" forces (which mediate the interactions within atomic nuclei).

The "Aether Electromagnetic System" is the "Medium" of "Light" and empty space.

Energy is electrical pulses of "Light" or photons that makes up all "Matter".

When God said in the beginning "Let there be Light", he spoke into being the first particle of "Matter".

"Light" Is "The Fact of Everything"!!!

"Matter" is made of photons or "Light".

God used "Aether" and "Light" to make all living and nonliving things in the world and in the universe.

The "Big Bang" was not the beginning; but the beginning of creation begin with "Aether" and "Light" when God said "Let there be "Light".

All things are made of "Aether" and "Light".

The four states of "Matter", solids, liquids, gases, and plasma, are made of the "Electromagnetic Spectrum of Light".

"Matter" is electrical energies of "Light" or pulses of "Light" called photons.

"Matter" is energy and energy is "Light" this is a fact.

"Are all four states of "Matter" made of "Light"?

'Yes, all four states of "Matter", solids, liquids, gases, and plasma are made of "Light".

All four states of "Matter" entails all that is considered mass and "Matter" in the entire universe.

The four states of "Matter" are entirely made up of the "Electromagnetic Spectrum of "Light".

Everything living and nonliving are made of the Electromagnetic Spectrum of "Light", and therefore everything is made of "Light".

Solids are "Matter", liquids are "Matter", gases are "Matter", plasma is "Matter", and "Matter" is "Light".

"Matter" and "Energy" and "Light" are all one and the same.

Notice the circles in this photo, of a light bulb radiating with the "Aether Electromagnetic System"

"Does "Light" propagate or travel in a vacuum?

As far back as the 1800s, people such as Sir Isaac Newton, Michael Faraday, James Clerk Maxwell, Nikola Tesla, and nearly everyone in the scientific community believed that "Light" like sound and air, and waves and water, they all needed a "Medium" in order to propagate or travel through space.

The famous Michelson-Morley experiments of 1887, was a failed attempt to prove that "Light" travels through a "Medium" of "Aether".

In the year 1905 Albert Einstein convinced the scientific community that "Light" did not need a "Medium" to travel and propagate through space.

It is the scientific community's belief today in 2017, that "Light" travels in the vacuum of space, and no "Aether of Light" is needed, and nor does it exist.

Einstein convinced the scientific community with his theory of "Special Relativity" and his study of the speed of "Light" (186,282 miles per second), that "Light" travels in a vacuum and did not need a "Medium".

It should be noted that Einstein did not actually PROVE the constancy of the speed of light in all frames of reference. Rather, it is an axiom (an underlying assumption) from which he derived the rest of his theory. The axiom can be experimentally verified, but it is not proven in any theoretic sense.

Until Randy Lee Holmes discovery of the "Aether Electromagnetic System" in 2016, there was no visible proof that the "Aether" actually existed.

He discovered "Aether" by shining different kinds of "Light" onto different dark screens of plasma televisions.

He noticed that every candlelight flame, match flame, AC light bulb, DC bulb, and every form of "Light" when placed in front of the dark television screen had the same "Aether Electromagnetic System Blueprint" accompanying each individual beam of "Light".

He also noticed that each beam of "Light" would move in the direction the flashlight was pointed, but the "Aether Electromagnetic Radiation Blueprint" remained stationary on the screen at all times.

Randy Holmes took more than 200 photos and 46 videos of this "Aether Electromagnetic System" and it's "Aether Electromagnetic Radiation Blueprint".

The discovery of "Aether" means that "Light" can travel faster than the known speed of "Light" (186,282 miles per second).

The discovery of "Aether" will bring about many applications in technology.

To name a few, Anti-gravity spacecraft, Force fields to protect our person, and protect our property, Buildings that levitate in the sky, Teleportation of human beings and objects, Travel to other planets in seconds, Travel to other dimensions, and the applications are limitless.

"Dose "Light" need a "Medium" to propagate?

People think that "Light" travels through the vacuum of space without a "Medium" because they don't know any better.

But, even the simple of minds should know that "Light" is an Electromagnetic Transverse Wave that consist of an electrical and magnetic field.

This electrical field that accompanies all "Light" is the "Aether" or "Medium" of "Light".

James Clerk Maxwell proved that any electrical field must also have a magnetic field.

Actually, Michael Faraday made the discovery and Maxwell proved it with his mathematical equations.

Without this "Aether" or "Electrical" and "Magnetic" field of the transverse wave, "Light" cannot propagate or travel in empty space of a vacuum.

Today the whole world shall see the "Secret of the universe" in the following two sentences.

"Light Is the Fact of Everything!!!

EVERY PHOTON Of "LIGHT" CARRIES AN INDIVIDUAL "AETHER ELECTRICAL AND MAGNETIC FIELD", or antenna that permeates and propagates throughout the universe; and beyond.

So to put it simply, if you have "Light" you will have "Aether", because one cannot exist without the other.

Example:

Many physicists wonder if "Aether" exists in our atmosphere.

I say to you...Do we not have sunlight?

The "Aether" signature and fingerprint of a light in my living room.

"Does all "Light" travel by way of "Aether"?

Yes, each photon of "Light" has its own individual "Aether" or electrical magnetic antenna that serves as a "Medium" for the propagation or travel of "Light".

"Aether" is the "Electrical" and "Magnetism" component of the transverse wave of "Light" that enables the wave to travel throughout the universe and through time and space.

"Light" to most of us is Electromagnetic Radiation seen as a self-propagating transverse oscillating wave of electrical and magnetic fields.

"Light" in reality is an "Electromagnetic Photon" that propagates or travels by way of individual Electromagnetic "Aether" as its "Medium" to permeate throughout the universe.

The top light is a flashlight and the bottom is a match flame.

"Is "Light" both a particle and a wave?

Einstein believed **light** is a **particle** (photon) and the flow of photons is a **wave**.

Research and experiments have proven this to be true.

We know that "Matter" is made of electrical components such as electrons, protons, and neutrons.

These components are all made of "Light" or photons.

Einstein in 1905 was awarded the Nobel Prize for his discovery of the photoelectric effect.

Now with the discovery of the "Aether Electromagnetic System" (AEMS), we can pursue many applications by using both the particle and the wave aspects of "Light".

The particle and wave aspects of "Light" will provide the possibilities of applications such as Anti-gravity spacecraft, Force fields to protect our person and our property, Buildings that levitate in the sky, Teleportation of human beings and objects, Travel to other planets in seconds, Travel to other dimensions, and the applications are limitless.

"Is all "Light" made from the Electromagnetic Spectrum"?

Yes, the Electromagnetic (EM) Spectrum consist of all the types of "Light" Radiation throughout the entire universe.

What is important to know is that all types of Electromagnetic "Light" is what makes up every particle of "Matter".

"Matter" is Electromagnetic Radiation of "Light".

"Matter" is what all living and nonliving things are made of; and "Matter" is "Light".

The "Electrical Magnetism" or "Aether" of "Light" is what bonds us electrically to all living and nonliving things made of "Matter" or "Light".

As we have discovered earlier in this book that we are all made of "Light" or "Matter".

We also know that "Matter" and "Light" and "Energy" are all one and the same.

There is no form of "Light" outside the Electromagnetic Spectrum.

All "Light" in the world and the universe exist within the Electromagnetic Spectrum.

"Did God create the world and the universe when He said "Let there be Light"?

On the first day of creation God said "Let there be "Light" and on the fourth day God created the sun.

Yes, God was the "Light" before the sun.

When God said "Let there be Light", He started at that moment to create all living and nonliving things with "Aether" and "Light".

God did not create one without the other because together this "Electrical Magnetism" Aether of Light" could combine the living to the nonliving as one energy.

This is how we are connected to all "Light" and "Matter" of the universe.

All "Matter" is made of "Light"; and "Energy".

If you read the Bible you will see many scriptures referring to God and Jesus as "Light"; and how we were created in His image.

The world and the universe is all one entity of electrical energy connecting all Electromagnetic Spectrum of Radiation together to form one system of "Light Energies".

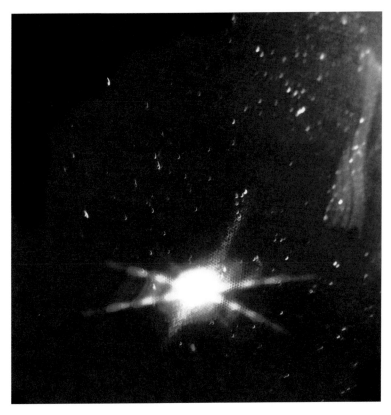

"Is it possible to travel faster than the speed of "Light"?

Einstein said nothing can travel faster than the speed of "Light" of 186,282 miles per second.

Einstein did not believe in the "Aether", and he convinced the world in 1905 that "Light" traveled in the vacuum of space and did not need a "Medium".

Einstein later in 1920 made a public announcement that he had made a mistake in early findings, and he stated that "General Relativity" and "Special Relativity" could not exist without an "Aether".

In 2016 retired high school teacher Randy Lee Holmes of Fayetteville, North Carolina, USA, discovered "Aether" in the form of an "Aether Electromagnetic Radiation System Blueprint".

Holmes discovered visual proof that each photon of "Light" came with its own individual "Electromagnetic Aether Antenna" that permeated throughout space-time and the universe.

By using a black plasma television screen he was able to take photos and videos of "Aether" and the "Aether Electromagnetic Radiation Blueprint" that accompanied every photon particle of "Light".

With the discovery of "Aether", it is now possibly to do experiments that require faster than the speed of "Light".

With the knowledge and proof of the existence of "Aether" we can now pursue limitless applications at the quantum level of physics.

Man shall soon learn the "Wheel-work" of the universe (Tesla's vision).

"Can Time and Space exist without Light"?

No, because "Matter" is "Electrical Magnetism Energy" that makes up Time and Space.

To put it simply, "Matter" is Time and Space or Space-Time.

Time and Space is part of the "Aether Electromagnetic System" (AEMS) that is connected to "Matter" beyond the quantum level of existence.

This is how the "Aether Electromagnetic System" will be used to travel to other planets, and other dimensions in time.

The "Aether Electromagnetic System" is a system of "One Electromagnetic Energy" that is connected to each photon of the universe, and beyond.

This "One Electromagnetic Energy" will allow limitless applications, and does exist as the most powerful energy of the universe, and beyond.

"What is Gravity"?

"Gravity" is the "Magnetism" component of the "Aether Electromagnetic System" (AEMS).

"Magnetism" of the "Aether Electromagnetic System" is what holds "Everything" or "Matter" together; you call it gravity.

The Higgs boson particle does not give everything mass and, in turn, holds the universe together.

"Magnetism" gives "Everything" mass and holds the universe together.

The "Aether Electromagnetic System" or "Aether" is a system that operates as the "Wheel-work" of living and nonliving "Matter" throughout the entire universe, and beyond.

Until the discovery of the "Aether Electromagnetic System" in 2016, no scientist, no physicist, no astrophysicist, and no one knew what "Gravity" was of the four fundamental forces of nature.

The best that the physics community could come up with was the "Graviton"; sadly to say.

The discovery of the "Aether Electromagnetic System" will start a new age of Anti-gravity technology.

The economy of the world runs on oil, so for many this is not good news; but all civilizations must make way for advancements in technology, or eventually face extinction.

The year 2017 shall be known as the year that man discovered "The Secrets of the Universe".

This discovery of the "Aether Electromagnetic System" (AEMS) will revolutionize the world of physics.

The greatest thing of all is that we will be a part of this exciting and historic age; as it came about during our lifetime here on earth.

What is "Dark Matter"?

"Dark Matter" is "Matter" that the human eye cannot see.

It is actual "Matter" of the universe.

"Light" is "Matter" held together by "Magnetism" of the "Aether Electromagnetic System", and forms "Gravity".

"Gravity" uses "Magnetism Light" to give form to objects for the human eye to see.

"Matter" is an "Electrical Entity" that has an electrical field emanating from it, even in its "Matter" ground state of existence.

This emanating electrical field that surrounds all "Matter" is "Dark Matter"; "Matter" that the human eye cannot see; it is an electrical field of "Matter" in its ground or resting state.

The universe is expanding because all "Matter" is surrounded with its electrical field in its ground state called "Dark Matter"; and same fields repel.

All "Matter" and it's "Dark Matter" electrical field in the universe carries the same resting charge, and the same charges repel one another, and expand away from one another.

"Dark Matter" is the electrical field of "Matter" that surrounds and emanates from all "Matter".

When "Light" is bending around objects it is traveling around or over "Dark Matter" to reach its destination.

When Einstein did his 1919 eclipse experiment he proved that "Light" would bend around massive objects like the sun "Light" was able to bend around the moon.

The "Light" was bending around the moon's "Dark Matter".

"Dark Matter" makes up 23% of all mass and energy in the entire universe.

Randy Lee Holmes has done several experiments in 2003 that proves "Dark Matter" existence, and the repelling and expansion of "Matter" when two objects of "Matter" comes in contact with each other. Ordinary "Matter" only makes up 5% of "Matter" in the entire universe.

What is "Dark Energy"?

"Dark Energy" is actually "Light Energy" that does not want to become mass.

One can look at it as "Electrical Light" that has not undergone "Magnetism" to form "Gravity" to form "Matter".

"Dark Energy" is the "Medium" of electrical and magnetism energy provided by the "Aether Electromagnetic System" (AEMS) of empty space.

"Dark Energy" makes up 72% of the total energy in the entire universe.

"Dark Energy" is the "Aether" of the universe.

The "Magnetism" of the "Aether Electromagnetic System" forms "Gravity" and "Gravity" forms "Matter".

"Dark Energy" lacks "Magnetism", "Gravity", and "Matter", but is the "Electrical Energies" that permeates empty space.

"Dark Energy" is the "Electrical Energies" of the "Aether Electromagnetic System" (AEMS) that permeates the universe, and beyond, and bonds living and nonliving things together as one energy.

Yes, you could one day use your mind to transport yourself to another planet, another universe, another dimension.

"Cold Fusion"

"Cold Fusion" means to fuse two atoms together at close to room temperature.

This is 2017, and to date, the physicist community will tell you that this has never been done.

It takes a huge amount of energy to fuse two atoms together without using extreme hot temperatures because of the repulsive forces of an atom.

If you could get the two atoms close enough the nuclear strong force would overwhelm the electric repulsive force, and slam them together (releasing far more energy than initially went into forcing them together).

Randy Holmes built the world's first "Cold Fusion" manually operated machine in 2003.

This machine would fuse the two electrons together of plastic and nylon fabric during an electrostatic interaction.

While pressing the top plastic and pulling the nylon fabric across the surface of a sheet of stationary plastic mounted on a hard plastic insulator, he was able to capture "Cold Fusion" in the top plastic sheet that hard pressed the nylon fabric.

The top plastic sheet acted as a capacitor which allowed the transport of the "Anti-gravity Cold Fusion Charge" to any location.

Place the top sheet of plastic anywhere on the inside or outside, on the floor, or on the ground, small objects and dead insects would float in midair on top of it.

Randy Holmes took the top sheet to school and showed many teachers and students how small objects can float in midair, before he retired from teaching in 2016.

The "Anti-gravity Cold Fusion Charge" would only last fourteen minutes at a time, depending on the amount of charge and the atmosphere.

The charge would maintain 14 minutes and thereafter still holding 20% of its beginning charge; after which the top sheet of plastic would need recharging again.

Randy Holmes can only generate charges of "Cold Fusion Anti-gravity Charge" large enough to levitate small objects at this time.

But, in the near future he hopes to generate "Cold Fusion Charges" large enough to float anti-gravity spacecraft that will provide the world with anti-gravity transportation.

This same "Cold Fusion Charge" will allow applications of Anti-gravity Force Fields, Buildings levitating in midair, and many others.

Photo of DC Battery Flashbulb from bulb of camera.

"Aether" photo taken with flashbulb of camera. Photo of the "Aether Electromagnetic System" (AEMS) Photo of the "Electromagnetic Radiation Blueprint" of all "Light".

Photo of DC Battery flashlight while standing to far right. White light middle
light is leaning to right.

Torch flame fire at top. Flashbulb shows the electromagnetic spectrum colors inside the triangle on the left; but outside the triangle on the right.

Also notice the middle white light line and the perpendicular white light line.

DC battery flashlight with photo taken while standing to far right of "Aether" fingerprint or radiation blue print.

Also reflection light at very bottom.

DC battery flashlight.

This middle white light line is leaning to the left because I am standing to far left of screen.

Flashbulb from camera at top flame from torch at bottom.

(AEMS)

Taken with flashbulb of camera on top.

Flame torch of fire on bottom.

Reflection light of flashbulb on very bottom.

Fire torch flame or fire at the top.

DC battery flashlight at the bottom.

Two very small lights at the very bottom.

Flame torch of flame fire on top.

Flashbulb from camera on bottom.

Small reflection light at very bottom of torch flame.

"Aether Radiation Blueprint" of flashbulb from camera at the top.

Candle light flame or fire on the bottom.

DC flashlight notice the color bands are on the inside of left triangle, and color bands are on the outside of triangle on the right.

AC light bulb with reflection on bottom. Beautiful cross, and beautiful "Aether Electromagnetic Radiation Blueprint", or "Radiation Signature", or "Electromagnetic Radiation Fingerprint".

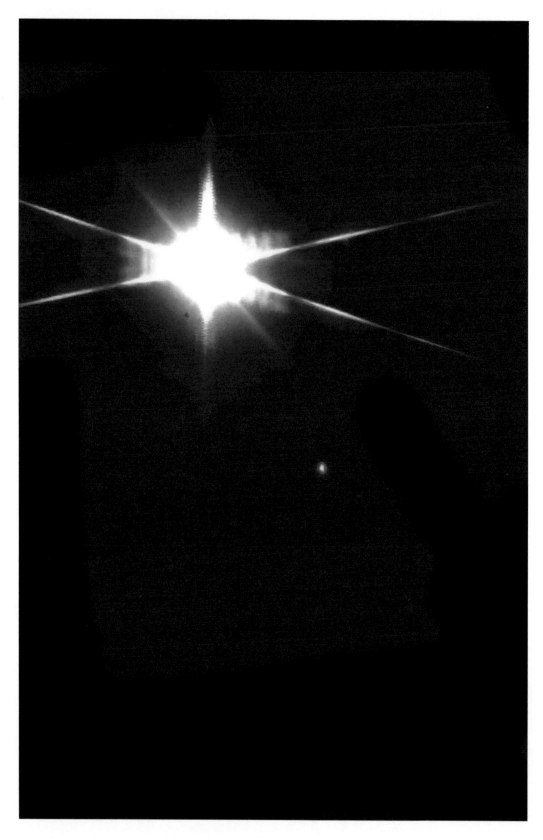

Top is flashlight. Bottom small (AEMS) is match flame fire.

The "Aether" signature and fingerprint of a DC light in living room.

Notice the beautiful color bands of the electromagnetic spectrum. You are looking at a star in the cosmos and its "Electromagnetic Radiation Blueprint".

(AEMS)

(AEMS)

(AEMS)

(AEMS)

(AEMS)

(AEMS)

(AEMS)

(AEMS)

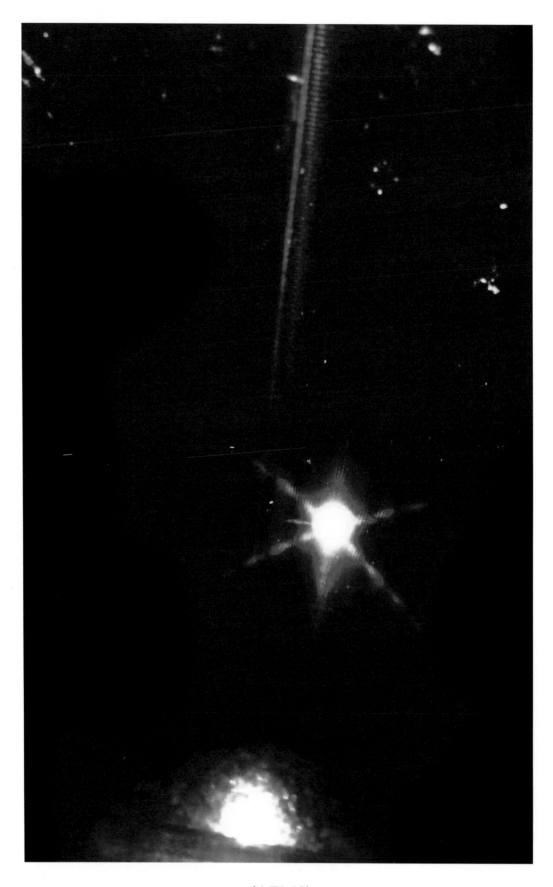

(AEMS)

(AEMS)

(AEMS)

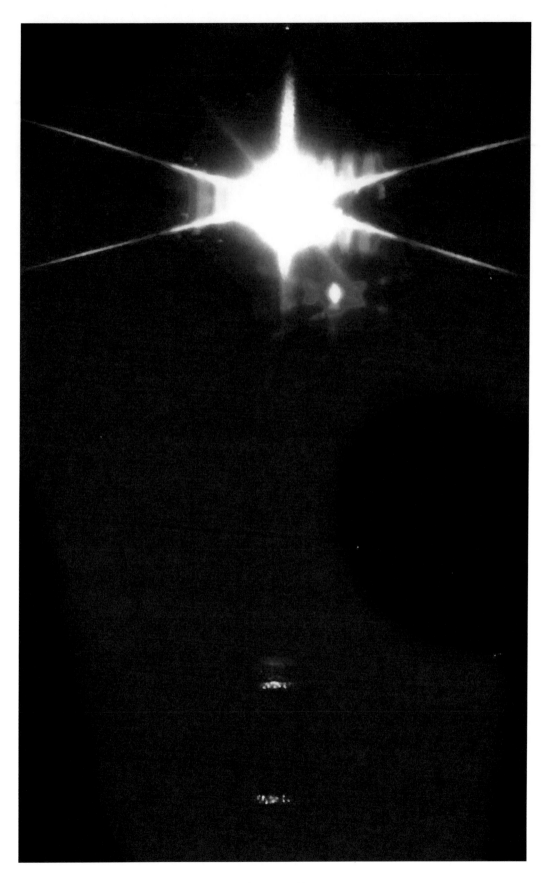

(AEMS)

(AEMS)

(AEMS)

Only one man knows in the entire world.

1. Where did we come from?

2. What is "Gravity"?

3. What is "Matter"?

4. What is "Dark Matter"?

5. What is "Dark Energy"?

6. What is "Aether"?

7. What is "Light"?

After reading his book and finding out the true secrets of the universe, many are forced to wonder if this is indeed the greatest discovery of our time.

Take the "Light" IQ Test

Only for the smartest people in the world!!!

1. Are all human beings and all things made of light?

2. Is matter and light both the same?

3. Are all four states of matter made of light?

4. Does light propagate or travel in a vacuum of empty space?

5. Does light need a medium to propagate or travel?

6. Does all light travel by way of Aether?

7. Is light both a particle and a wave?

8. Does light travel by way of waves?

9. Is all light made from the Electromagnetic Spectrum?

10. Is each individual photon created as an Electromagnetic Wave?

11. Is it possible for light to travel faster than 186,282 miles per second?

12. Can time and space exist without light?

13. At the quantum level is all matter made of photons?

14. Can light exist before the sun exist?

15. Is electricity combine with magnetism the same as Aether?

16. Is matter and energy both the same?

17. Can light exist without Aether?

18. Are all things living and nonliving powered by Aether?

19. Did God use light and Aether to create all in the world and in the universe when He said "Let There Be Light"?

20. Is "Aether" responsible for "Gravity"?

Answer key:

5 times 20 = 100% 100% = Perfect Score of 1000.

Example to calculate score:

20% right = score of 200

30% right = score of 300

90% right = score of 900

1y2y3y4n5y6y7y8y9y10y11y12n13y14y15y16y17n18y19y20y

Only .001% of the human population can score 1000 on this test.

Please do not consider yourself dumb if you do not score 300 or less on this test.... just do some research and become smarter.

The average score is 5 of 20 questions correct or 25% or score 250.

This book was written and Copyrighted by Randy Lee Holmes 2016.

Printed in the United States
By Bookmasters